# STAND UP

## Against Climate Change

Stuart A. Kallen

ReferencePoint
Press®

San Diego, CA

© 2022 ReferencePoint Press, Inc.
Printed in the United States

**For more information, contact:**
ReferencePoint Press, Inc.
PO Box 27779
San Diego, CA 92198
www.ReferencePointPress.com

LIBRARY OF CONGRESS CATALOGING-IN-PUBLICATION DATA

Names: Kallen, Stuart A., 1955-, author.
Title: Stand up against climate change / Stuart A. Kallen.
Description: San Diego, CA : ReferencePoint Press, 2021. | Series: Teen
   activism library | Includes bibliographical references and index.
Identifiers: LCCN 2021009651 (print) | LCCN 2021009652 (ebook) | ISBN
   9781678201463 (library binding) | ISBN 9781678201470 (ebook)
Subjects: LCSH: Climatic changes--Juvenile literature. | Climate change
   mitigation--Juvenile literature. | Environmentalism--United
   States--Juvenile literature. | Environmentalists--United
   States--Juvenile literature.
Classification: LCC QC903.15 .K34 2021  (print) | LCC QC903.15  (ebook) |
   DDC 363.738/740973--dc23
LC record available at https://lccn.loc.gov/2021009651
LC ebook record available at https://lccn.loc.gov/2021009652

# CONTENTS

# Climate Change Is Here and Now

Environmental activist Xiye Bastida was born in the town of San Pedro Tultepec, Mexico, in 2002. Her father is a member of the indigenous Otomi tribe. When Bastida was growing up, she says she learned important lessons from the Otomi: "Earth is our home. It gives you air, water and shelter. Everything we need. All [the earth] asks is that we protect it."[1] In 2014 Bastida saw what can happen when the earth is not protected. Her town experienced a prolonged drought made worse by the effects of climate change. When a local lake dried up for the first time in memory, the lives of farmers and fishers in the region were badly disrupted.

The three-year drought in San Pedro Tultepec was followed by a year of extreme rain and flooding. This forced Bastida's family to leave their home and move to New York City in 2015. But climate change remained on Bastida's mind when she saw what Hurricane Sandy had done to New York. Although the deadly superstorm had hit in 2012, the region was still recovering. Scientists blamed the changing climate for the extreme destruction wrought by Sandy. The drought in Mexico and the hurricane in New York moved Bastida to become a climate activist. As she explains, "Wherever you are, the climate crisis is affecting everyone, everywhere."[2]

In 2018 Bastida joined Fridays for Future (FFF), founded by fifteen-year-old Swedish climate activist Greta Thunberg. The movement coordinates climate strikes; students walk out of school on select Fridays, demanding that political leaders deal immediately with the climate emergency. In March 2019 Bastida became an FFF leader. She organized a strike at her New York high school that was joined by six hundred students.

"Wherever you are, the climate crisis is affecting everyone, everywhere."[2]

—Xiye Bastida, environmental activist

The FFF movement grew into a worldwide movement called Global Week for Future. Bastida organized a youth activism training program to help students prepare for a September 2019 strike. Around 4 million people in 150 countries participated in the Global Week for Future strike. At the time, this was the largest climate protest in history. Thunberg appeared at the New York

*Environmental activist Xiye Bastida addresses people at a rally in front of City Hall in New York City. She became concerned about climate change as a child when her hometown in Mexico experienced a prolonged drought due to climate change.*

City demonstration, speaking to a crowd of more than 250,000: "We will do everything in our power to stop this crisis from getting worse, even if it means skipping school or work, because this is more important. Why should we study for a future that is being taken away from us?"[3]

## A New Narrative

Bastida is one of countless teenagers around the world who are organizing, marching, and raising their voices to draw attention to the climate crisis. According to a 2019 *Washington Post*–Kaiser Family Foundation poll, one in four American teenagers is taking action against climate change. Some have participated in FFF, while others have joined climate action groups such as the People's Climate Movement, the Sunrise Movement, and the Extinction Rebellion. A growing number of activists are, like Bastida, members of indigenous or marginalized communities. As twenty-one-year-old Philippine activist Mitzi Tan explains, "Vulnerable communities should be leading the movement on the climate crisis, because we're the ones most affected by it."[4]

While young activists have been talking about climate change for decades, the newest generation is getting more attention in the media. Some of the credit goes to Thunberg's activism, but there are other elements at work. Today's protesters are coordinating their activities—and amplifying their voices—through social media. Videos of young climate protesters confronting politicians or showing the aftereffects of climate disasters have gone viral. This online visibility has drawn even more young people into the movement, which fuels its growth. Bastida speaks for many young climate advocates when she explains what drives her activism: "We need to change our culture and change our narrative. For too long, the narrative has been that this is some big distant thing that will happen in the year 2100. But pollution is here. Heatwaves are here. Wildfires are here. Melting ice caps are here. Floods are here. Category 5 hurricanes are here. It's here already."[5]

# CHAPTER ONE

# The Issue Is Climate Change

In January 2021 climate scientists at the University of Leeds in England released an alarming report. Research revealed that the ice on earth was melting much faster than current climate models predicted. The Leeds study looked at ice that was once considered a permanent part of the earth's environment. The ice includes massive sheets covering the Arctic Ocean, around two hundred thousand mountain glaciers, and vast expanses of frozen water more than 1 mile (1.6 km) thick at the North and South Poles.

Researchers examined ice melts from 1994 to 2017. They found that during the 1990s the earth was losing an average of 760 billion tons (689 billion metric tons) of ice per year. By 2017 that number had grown to more than 1.2 trillion tons (1.09 trillion metric tons). That amount of ice is roughly equivalent to a sheet of frozen water 328 feet (100 m) thick covering the entire state of Michigan. While the numbers are almost impossible to grasp, the conclusion of the study is not difficult to understand. According to ice sheet researcher Robin Bell, "Ice on our planet is melting. We have turned up the temperature, and just like you can watch an ice cube in your glass melt on a hot summer day, our actions are melting our planet's ice."[6]

The actions Bell refers to include driving gas-powered vehicles, using coal to generate electric power, and burning natural gas to produce everything from food to steel and cement. In other words, the ice is melting due to modern society's consumption of fossil fuels.

## Planet-Warming Problems

Coal, oil, and natural gas have contributed to the rapid advancement of human society and culture since the nineteenth century. Fossil fuels directly or indirectly power much of the world's transportation, refrigeration, heating, lighting, water purification, and medical and telecommunications equipment. But all those modern conveniences come at a cost; fossil fuel consumption is causing the planet to get hotter. This is due to the greenhouse effect: when fossil fuels are burned, they produce gases that trap the sun's heat in the atmosphere.

When coal, oil, and natural gas are burned, they produce carbon dioxide ($CO_2$). This gas is responsible for about 80 percent of planet-warming pollution. Methane, emitted by livestock and through fossil fuel production, accounts for around 10 percent of climate pollution. Nitrous oxide, used to treat wastewater and produce fertilizer, is responsible for roughly another 7 percent of all planet-warming gases. Since 1900 these gases in the atmosphere have caused average global temperature to increase by about 2°F (1.1°C). While two degrees does not sound like much, the rising temperatures have upset the balance of nature. The rising heat is behind the extreme droughts and massive wildfires that have disrupted life from the American West to Australia. In the Midwest, climate change is increasing the number and intensity of destructive rainstorms and floods. The oceans, which absorb around 90 percent of the heat trapped in the atmosphere, are also warming. Ocean surface temperatures have increased by an average of 1°F (0.56°C) over the past century. This has fueled bigger and deadlier hurricanes and tropical cyclones from Florida to the Philippines.

The problems caused by climate change are expected to increase as the planet continues to warm. The five hottest years on record have all occurred since 2015. Average summer temperatures in 2020 were the highest ever recorded around the globe, according to the US science agency National Oceanic and Atmospheric Administration. At least fifty cities worldwide set records for extreme heat as temperatures rose to record levels on every continent. And by 2040 the number of heat waves is expected to quadruple worldwide unless carbon emissions are drastically reduced, according to a study by the journal *Environmental Research Letters*.

## Warning the World

In the 1980s, when environmental activists first started talking about climate change, the issue was referred to as global warming or the greenhouse effect. The problem first made international headlines in 1988 when climate scientist James Hansen appeared before the US Senate Committee on Energy and Natural Resources. Hansen bluntly warned senators that there was a 99 percent certainty that humanity was causing the climate to change

This sign in the Kenai Peninsula, Alaska, indicates how far the glacier that appears in the distance has retreated since the year 1978. This massive glacial retreat is typical as the earth's climate warms.

by burning fossil fuels: "It's time to stop waffling so much and say that the greenhouse effect is here and affecting our climate now."[7]

Before Hansen's Senate testimony, fewer than 40 percent of Americans had ever heard of climate change. Most who had heard about global warming thought it was a problem for future generations. But in 1988, after Hansen's words made headlines around the globe, 58 percent of Americans said they had heard about global warming.

After Hansen's testimony, research on climate science increased significantly. Climatologists ran thousands of increasingly sophisticated experiments to prove how and why the climate was changing. In 2020 the National Aeronautics and Space Administration (NASA) revisited seventeen research projects conducted over a thirty-year period. The predictions in fourteen of the studies proved to be very accurate when compared to actual changes in the atmosphere. The studies neither overestimated nor underestimated the actual observed warming of surface and ocean temperatures. As director of NASA's Goddard Institute for Space Studies Gavin Schmidt said in 2020, "The results of this study of past climate models bolster scientists' confidence that both [the older studies] as well as today's more advanced climate models are skillfully projecting global warming."[8]

## Science and Politics

Despite the scientific consensus, climate change became a polarizing issue over the years. Even in 2021 millions of Americans either doubted that climate change is serious or denied that the problem exists at all. This has left activists to fight more than climate change; they also must push back against climate change deniers. To understand why, it is helpful to look at how the science of climate change was politicized.

Almost immediately after Hansen testified before the Senate, major emitters of global warming pollution sought to cast doubt on the severity of climate change. Multinational oil companies, coal producers, steelmakers, and other polluters hired research-

Most of the ice on earth is located on land in Greenland, the Arctic Circle, and Antarctica. When the ice melts, the water runs into the oceans, raising sea levels around the globe. According to the United Nations' Intergovernmental Panel on Climate Change, if ice continues to melt at its current rapid rate, sea levels will rise by as much as 3 feet (1 m) by 2100. This would make dozens of coastal cities uninhabitable, including New York City, Miami, Venice, Italy, Bangkok, Thailand, and Hong Kong.

Rising sea levels are already causing floods and other problems in many places. And as the ice melts, the earth warms faster because ice plays a very important role in regulating the climate. When the sun's rays hit the earth, 70 percent of the heat is absorbed by the land, oceans, plants, and objects like buildings and roads. Around 30 percent of the sun's energy strikes ice and snow at the poles and elsewhere. This reflects the heat of the sun back into space. With the ice at the poles melting faster than ever, the Arctic is warming twice as fast as the rest of the planet.

ers to cast doubt on the work of climate scientists. This small but vocal group of researchers conducted industry-financed studies that either denied temperatures were rising or falsely blamed global warming on sunspots or long-term weather patterns that had nothing to do with human activity.

While climate deniers published industry-funded studies, wealthy conservatives and major corporations spent millions of dollars to found policy institutes, or think tanks, whose main purpose was to cast doubt on climate science. Think tanks like the Competitive Enterprise Institute and the Heartland Institute, which continue to operate, produce climate-denial editorials in the media and provide spokespersons to voice their opinions on news programs and talk shows. As one unnamed historian stated, this type of debate was "transforming the issue from one of scientific concern to one of political controversy."[9]

In addition to funding climate-denial think tanks, corporations have funded the political campaigns of antiscience politicians. In 2019 there were 150 Republican members of Congress who denied the scientific consensus that climate change is a threat to humanity. Together these climate deniers received more than $68 million in direct contributions from the fossil fuel industry, according to the Center for Responsive Politics, which tracks political donations.

The efforts of the deniers succeeded in turning climate change into a political wedge issue. In 2020 polling from the Yale Program on Climate Change Communications showed that 18 percent of Americans were doubtful or dismissive of climate change science, while 8 percent were disengaged or uninterested in the problem.

## Climate Conventions

While Americans might be divided on the issue of climate change, the international community has responded. In 1988 the United Nations and the World Meteorological Organization founded the Intergovernmental Panel on Climate Change (IPCC) to provide unbiased scientific information about human-induced climate change. The IPCC is composed of twenty-five hundred experts from more than sixty countries who work in widely divergent fields such as climatology, ecology, economics, medicine, and oceanography. In 1990 the IPCC released its first climate report, which concluded, "We are certain emissions resulting from human activi-

As automobiles burn fossil fuels, they emit carbon dioxide, which traps the sun's heat in Earth's atmosphere, causing rising temperatures.

ties are substantially increasing the atmospheric concentrations of the greenhouse gases: carbon dioxide, methane . . . and nitrous oxide. These increases will enhance the greenhouse effect, resulting on average in an additional warming of the Earth's surface."[10]

The IPCC released another report with similar conclusions in January 1992. The following June the United Nations Conference on Environmental Development held a twelve-day convention in Rio de Janeiro known as the Earth Summit. The Earth Summit was attended by heads of state from 102 countries. About seventeen thousand others from environmental groups and nongovernmental organizations also attended the summit.

Delegates at the Earth Summit produced the United Nations Framework Convention on Climate Change (UNFCCC), a treaty that was signed within months by government leaders in 154 countries, including the United States. The UNFCCC stated that industrial nations should take immediate steps to reduce climate-changing pollution. However, the framework was completely voluntary. Signatories did not have to set specific goals for reduction, there was no time frame set for reducing emissions, and no penalties were proposed for those that violated the recommendations.

The Convention on Climate Change continued to hold annual conferences throughout the world. New goals to reduce climate pollution are set every year as the amount of $CO_2$ in the atmosphere continues to increase and the climate rapidly warms. One of the most substantial actions undertaken by the convention is known as the Paris Agreement. This international accord was signed by 197 countries. Signatories promise to substantially reduce greenhouse gas emissions to limit global warming. To do so, carbon pollution must be decreased by 45 percent by 2030 and reduced to zero by 2050. This would limit global warming to 2.7°F (1.5°C) by the end of the century.

## Shifting Priorities

The United States signed on to the Paris Agreement in 2015 when Barack Obama was president. But the political back-and-forth

When Joe Biden was elected president in 2020, he pledged to address the problems of climate change and environmental justice. The latter became part of his agenda thanks to the efforts of young activists who sought greater awareness of the effects of climate change on communities of color. Robert Bullard, leader of the Black Environmental Justice Network, explains, "It used to be that these mostly white, mainstream environmental groups would be . . . making the decisions and then call us to say what was decided. . . . We said, 'Never again. We are not going to leave it to other folks to speak for us.'"

The work of climate justice groups like the Sunrise Movement could be seen in Biden's choice to lead the US Department of the Interior, which manages 500 million acres (202 million ha) of land in the United States. Biden initially wanted to appoint a longtime Washington insider to head the department. However, after activists lobbied Biden, he nominated Debra Haaland, a US representative from New Mexico, to lead the department. After Haaland was confirmed in March 2021 she became the first Native American to lead the US Department of the Interior. She has promised to fiercely defend the planet from further warming.

Quoted in Evan Halper and Anna M. Phillips, "Environmental Justice Groups Seize Moment," *Los Angeles Times*, January 31, 2021. https://enewspaper.latimes.com.

over the agreement demonstrates the political problems faced by climate activists. When Donald Trump ran for president in 2016, he repeatedly called climate change a hoax. After he won the election, Trump eliminated or weakened more than 130 regulations aimed at reducing the effects of climate change, according to Columbia University's Sabin Center for Climate Change Law. The Trump administration also enacted policies that accelerated the production of fossil fuels and withdrew the United States from the Paris Agreement at the earliest possible date, which was November 4, 2020.

The day before the United States left the Paris Agreement, the country held a presidential election, which Trump ultimately lost to Joe Biden. When Biden was sworn in as president in January 2021, one of his first official acts was to sign an executive order to reenter the agreement. And during the first week of his administration, Biden moved quickly to provide a massive federal response to climate change.

On January 27, 2021, Biden signed executive orders to pause production of oil and gas on federal lands and in federal waters. Environmental activists view this as an important step in reducing carbon pollution since about one-quarter of the country's carbon emissions come from coal, oil, and gas produced on lands managed by the US government. Biden also called for the creation of a Civilian Climate Corps that would put young Americans to work restoring public lands and waters, planting trees, improving access to parks, and tackling issues related to climate change.

Biden pledged that he would replace all vehicles owned by the federal government with electric vehicles. This is a massive undertaking; the federal fleet includes 50,000 passenger vans, 225,000 postal service vehicles, and over 200,000 passenger vehicles. Together they burn almost 400 million gallons (1.5 billion L) of gasoline annually. As Biden stated when announcing these measures, "We've already waited too long to deal with this climate crisis. We can't wait any longer. We see it with our own eyes, we feel it, we know it in our bones. And it's time to act."[11]

> "We've already waited too long to deal with this climate crisis. We can't wait any longer. . . . And it's time to act."[11]
>
> —Joe Biden, forty-sixth president of the United States

## Making Politicians Keep Their Promises

Many climate activists were happy about Biden's initial steps to address climate change. At the same time, they understood that there was much work to be done. One problem was the way Biden pursued the new policies. When a president signs an executive order, it guides administration policies in many important ways. However, executive orders can be repealed by the next president. In fact, many of Biden's actions were aimed at reversing Trump's executive orders on environmental policy.

Members of the Sunrise Movement, a youth-led environmental group, demonstrate in favor of legislation to address climate change. The Sunrise Movement has over eighty-seven thousand members across the country.

If activists want to implement permanent solutions to climate problems, they need to convince Congress to pass legislation. Laws enacted by Congress cannot be undone by executive order. To enact legislation, climate activists have to fight back against powerful fossil fuel companies and the politicians they endorse. This means persuading enough Americans to vote for representatives who support a green energy agenda. Activists also plan to keep up pressure on Biden to ensure that he follows through with his agenda. Evan Weber, political director of the youth-led Sunrise Movement environmental group, cautions, "Young [activists] learned that you can't take these promises at their face value and that the work does not end after an election. We're definitely gonna ask nicely at first and when demands aren't met, we'll change tones and tactics."[12]

Weber is sure his organization has the power to push politicians. The Sunrise Movement has over eighty-seven thousand members in more than 475 chapters throughout the country. Activists purchase political ads, rally outside the offices of their congressional representatives, and use social media to apply pressure when Congress votes on environmental issues. According to Weber, "There is an activated, energized, mobilized base of young people that really, really wants change and will reward politicians and defend politicians who are allies . . . and will punish and go after people who are standing in the way of progress."[13]

Activists in the Sunrise Movement and other environmental organizations judge progress by many measures, including how new policies will affect disadvantaged communities. This social justice element is important because climate-related disasters most often impact communities of color. As a 2019 study by the journal *Science* showed, the poorest counties in the United States—in Texas, Mississippi, and Florida—are most vulnerable to severe flooding, droughts, heat waves, and hurricanes. Most residents in these counties are Black or Hispanic.

## The End of an Era

While activists were heartened by the political developments of 2021, the climate fight is far from over. While Biden's actions signal the beginning of a new environmental era, politicians backed by the fossil fuel industry immediately voiced their opposition to

> "The Biden Administration took a series of coordinated actions that . . . may well mark the official beginning of the end of the fossil-fuel era."[14]
>
> —Bill McKibben, climate activist

the new measures. And the ice continued to melt. But there was some optimism from author and scholar Bill McKibben, who has been waging a nationwide fight against climate change since 2007. McKibben wrote on the day Biden signed the executive orders on climate change, "January 27th was the most remarkable day in the history of America's official response to the climate crisis. . . . In the course of a few hours, the Biden Administration took a series of coordinated actions that, considered together, may well mark the official beginning of the end of the fossil-fuel era."[14]

# CHAPTER TWO

# The Activists

Bill McKibben is a vocal environmentalist who has been focused on the issue of climate change for decades. In 1989 McKibben wrote *The End of Nature*. This was one of the first nonscientific books about climate change, and McKibben wrote it with hopes that it would prompt people to act. In 2007 McKibben was still working to educate people about the hazards of climate change. He founded the organization 350.org, which calls for all fossil fuel production to cease while society transitions to 100 percent renewable energy sources.

The name 350.org is based on the way carbon dioxide is measured in the atmosphere, as parts per million (ppm). Scientists believe that the safe upper limit for $CO_2$ in the atmosphere is 350 ppm. Before industrial societies began burning coal, oil, and gas in the eighteenth century, $CO_2$ concentrations in the atmosphere were around 280 ppm. Sustainability scientist Kimberly Nicholas explains that those are the conditions "on which civilization developed and to which life on earth is adapted."[15]

Although $CO_2$ levels were around 380 ppm when McKibben founded 350.org, he picked the number as a way to remind people of the importance of reducing carbon pollution to avoid climate disaster. But his efforts, and those of more mainstream environmental organizations, did not have much of an effect. In 2021 the concentration of $CO_2$ in the atmosphere was 415 ppm and rising.

## Striking for the Climate

A new approach to climate activism seemed necessary, and it came from an unlikely source. In August 2018 fifteen-year-old Swedish environmental activist Greta Thunberg initiated a one-person environmental protest. Thunberg skipped out of her ninth-grade classes to sit outside the Swedish parliament. She held a homemade sign that said, "School Strike for Climate."[16] A reporter asked Thunberg what she was doing. She replied that she was planning to cut school every Friday until Sweden passed laws that drastically reduced carbon emissions. Two weeks later Thunberg spoke at a climate rally in Stockholm. "If people knew what a nightmare scenario we will face if we don't keep global warming below 2°C," she said, "they wouldn't need to ask me why I'm on school strike outside parliament. Because if everyone knew how serious the situation is and how little is actually being done, everyone would come and sit down beside us."[17]

As it turned out, Thunberg's dire message would soon inspire people throughout the world to sit down and hold climate strikes. After the story of her one-person protest was described in a local paper, her social media account blew up. Thunberg's school strike movement, which she called Fridays for Future, went viral. Thunberg became an international sensation as her simple message galvanized the world.

> "If people knew what a nightmare scenario we will face if we don't keep global warming below 2°C, they wouldn't need to ask me why I'm on school strike."[17]
>
> —Greta Thunberg, climate activist

In January 2019 Thunberg was invited to the World Economic Forum in Davos, Switzerland, where rich and powerful leaders gathered to discuss global issues. Thunberg called her speech "Our House Is On Fire." In it, she asserted, "Adults keep saying: 'We owe it to the young people to give them hope.' But I don't want your hope. . . . I want you to panic. I want you to

feel the fear I feel every day. . . . I want you to act as if our house is on fire."[18] In March, Thunberg was nominated for a Nobel Peace Prize.

Thunberg's climate strike movement continued to grow. In September 2019 an estimated 6 million people of all ages and backgrounds joined in international climate strikes known as the Global Week for Future. In what has been called the biggest climate protest ever held, people took to the streets in more than six hundred cities in the United States, Europe, Africa, and the Middle East.

## Zero Hour Takes on Politicians

Several weeks before Thunberg held her first school strike, she took part in a Zero Hour demonstration in Sweden. Zero Hour was cofounded in Seattle by a sixteen-year-old Colombian-born student named Jamie Margolin. Thunberg and Margolin soon became friends through social media. Thunberg later recalled, "Before that I basically hadn't met any young person who seemed to care about the climate, the environment, or our future survival on the planet. . . . I remember feeling so alone, it seemed as if no one my age . . . wanted to make a difference—apart from people like Jamie Margolin."[19]

Margolin was moved to start Zero Hour in September 2017 as two simultaneous environmental disasters unfolded. Hurricane Maria devastated much of Puerto Rico, while massive wildfires in Canada filled the Seattle sky with choking smoke. According to Margolin, "Hurricane Maria—which was clearly a climate crisis yet the media acted like it wasn't climate caused [was a catalyst for Zero Hour]—and literally not being able to breathe clean air for two weeks."[20]

Zero Hour held its first national event in Washington, DC, in July 2018. For three days Zero Hour activists spoke to congressional representatives and lobbied them to reject campaign donations from carbon polluters. As protester Maeve Secor explained, fossil fuel companies "are basically paying off public officials with campaign funds to not pass climate legislation."[21] Other Zero Hour protests occurred in about twenty-five other cities, including London, Las Vegas, and Nairobi, Kenya.

As part of her advocacy, Margolin gives a lot of speeches. Before the coronavirus pandemic shut down schools in March 2020, Margolin traveled the country talking to students about climate change. The work can be difficult, and Margolin says she always starts with an apology to students: "I know this is unfair. I wish the future could be better than this. . . . I was [once] sitting in your seats, not knowing what to do."[22] But Margolin also provides a positive message that empowers students to take action. "We have to be protesting and lobbying," she says. "But we also have to be organizing in our communities; doing education work. We have to be addressing the issue in the courts. We can't pretend like we have the luxury of choosing one solution."[23]

## Hardworking Activists

In 2020 Margolin was extremely busy. She published a guide-book for young activists called *Youth to Power: Your Voice and How to Use It*. Margolin also often stays up late into the night to work on Zero Hour campaigns. This work involves collaborating with others on Zoom and answering dozens of texts and emails every day. The efforts required by Margolin on a single campaign exemplifies the hard work performed by leading climate activists around the globe. Zero Hour organized a six-city bus tour in the Midwest called #Vote4OurFuture. The effort was geared toward getting young people to vote in the upcoming 2020 presidential election. Margolin needed to approve graphics and content created by other activists for use in press releases, flyers, and social media campaigns. She also needed to contact young activists and media organizations in each city on the bus tour, while arranging promotional events with local environmental groups.

*Environmental activist Jamie Margolin speaks during a rally by youth activists and others in Seattle in 2018. Prior to the onset of the coronavirus pandemic in March 2020, Margolin traveled the country talking to students about climate change.*

Margolin says there are a lot of other things she would rather be doing, but she dedicates her time to climate activism because adults have been irresponsible on the issue for so long. And she is joined by others who feel the same way. Seventeen-year-old Jonah Gottlieb, who worked on the #Vote4OurFuture campaign, was moved to activism after barely surviving the deadly 2017 wildfire in Petaluma, California. To express his frustration over the political response to climate change, Gottlieb joined Zero Hour. He was soon working up to eighty hours a week as a climate activist. "We all have our little place that we've carved out at school," he states, "and then we get home and work until 2 or 3 in the morning. That's just the youth-activist life."[24]

In Margolin's case part of the youth-activist life is appearing before Congress. In 2019 she gave congressional testimony in support of the Green New Deal. This resolution is designed to move the United States to 100 percent renewable energy by 2029. The Green New Deal calls for overhauling

> "Experts agree there are multiple pathways to decarbonize the U.S. energy system and that doing so is technologically and economically viable."[25]
>
> —Jamie Margolin, founder of Zero Hour

transportation, housing, agriculture, forestry, and other systems to eliminate carbon pollution and slow climate change. But the Green New Deal is more than a climate resolution. The program, sponsored by New York representative Alexandria Ocasio-Cortez, addresses issues related to environmental justice. Part of the Green New Deal would focus on improving the air, water, and overall health of vulnerable communities that suffer most from pollution and climate change.

The Green New Deal was a massive policy package that had little chance of passage by a deeply divided Congress. While the political fate of the resolution remains hazy, Margolin expresses optimism. She attests, "The good news is that experts agree there are multiple pathways to decarbonize the U.S. energy system and that doing so is technologically and economically viable."[25]

## Climate Change and Black Justice

One of the cosponsors of the Green New Deal was Ilhan Omar, a congresswoman who represents Minneapolis. Omar was born in Somalia in 1982 and moved to the United States when she was thirteen. She was one of only three Muslims in Congress in 2021. Omar is widely known as a very progressive politician in political circles. But those who follow TikTok might be more familiar with Omar for the videos made by her daughter Isra Hirsi. The videos range from the silly—Hirsi trying to teach her mother a viral TikTok dance—to the serious, as when they show Omar's work in Congress. But Hirsi is more than a TikToker; she is a dedicated climate advocate and social activist who cofounded the US Youth Climate Strike, a group focused on bringing more kids of color into the climate movement and elevating them to leadership positions.

Hirsi, born in 2003, joined her high school environmental club when she was fifteen years old. The experience helped her understand how environmentalism intersects with the Black Lives Matter movement. Hirsi learned, for example, that Black people in Los Angeles are almost twice as likely as Whites to die from heatstroke during heat waves. And while periods of extreme heat might be natural, the number of heat waves has tripled in cities like Los Angeles in the past fifty years due to climate change. Heat waves kill society's most vulnerable people—the very young, the elderly, the poor, and those who have chronic medical conditions such as diabetes and respiratory illnesses. People who live in urban areas and lack access to air-conditioning are the most frequent victims of heatstroke, which kills around six hundred Americans every year.

## Creating Her Own Movement

Hirsi wanted to work on environmental justice issues in which climate change problems overlap with those of systemic racism. She joined a Minnesota statewide advocacy group whose members were mostly older White activists. But Hirsi says she felt marginalized by their subtle racism. She says, "I'm almost always the

## Greta Thunberg in Black and White

Climate activist Greta Thunberg made headlines for a 2019 speech delivered to the world's most powerful people at the World Economic Forum. In this excerpt, she clarifies the price of inaction:

> Now is not the time for speaking politely or focusing on what we can or cannot say. Now is the time to speak clearly. Solving the climate crisis is the greatest and most complex challenge that Homo sapiens have ever faced. The main solution, however, is so simple that even a small child can understand it.
>
> We have to stop our emissions of greenhouse gases. And either we do that or we don't.
>
> You say that nothing in life is black or white. But that is a lie. A very dangerous lie.
>
> Either we prevent a 1.5°C of warming or we don't. . . .
>
> Either we choose to go on as a civilization or we don't.
>
> That is as black or white as it gets.
>
> There are no grey areas when it comes to survival.
>
> Now we all have a choice.
>
> We can create transformational action that will safeguard the living conditions for future generations.
>
> Or we can continue with our business as usual and fail.
>
> That is up to you.

Greta Thunberg, *No One Is Too Small to Make a Difference*. New York: Penguin, 2019, pp. 29–30.

first young person that is black in these [groups]. . . . I learned a lot about how climate activism isn't meant for people like me. They don't have people like me, and in order to make spaces welcoming of black and brown voices, you have to create them yourself."[26]

This led to the creation of US Youth Climate Strike. During the run-up to the 2020 presidential election, Hirsi used the group to push for a debate that would focus solely on climate change. While Hirsi was not old enough to vote in the election, she was able to use her mother's political access to personally contact

Democratic politicians who were running for president. This group included Senators Bernie Sanders, Amy Klobuchar, and Elizabeth Warren. Hirsi's work paid off; CNN featured a debate on climate change that featured all the Democratic candidates. But taking advantage of her mother's connections was an unusual move for Hirsi, who likes to keep a low profile. "I don't like getting recognized," she says. "It doesn't happen a whole lot, but it does, and it makes me uncomfortable. . . . I also don't like the spectacle, the amount of judging, or watching down. I don't tweet as much as I used to because there [are] so many people watching."[27]

Some of those people were powerful Republicans, including then-president Donald Trump, who singled out Omar in a series of hateful tweets. This caused a surge of death threats directed at Omar, Hirsi, and other members of her family. The abuse on social media sometimes spilled over into Hirsi's accounts, in which classmates and total strangers posted vile messages. While the comments might have been devastating to many teenagers, Hirsi remained calm. She is used to dealing with security threats, as she wrote in a 2020 Medium essay: "I live 2 lives—that of a climate justice organizer and that of a politician's daughter."[28]

> "Understand that you are important and your voice is too, and don't let adults, especially, tell you otherwise."[29]
>
> —Isra Hirsi, founder of US Youth Climate Strike

After the Trump controversy, former president Barack Obama sent his support to Hirsi in a tweet. Obama praised her work as a climate activist—and did not mention her mother. This pleased Hirsi. While she is very proud of her mother, Hirsi remains laser focused on organizing young people to advocate for the climate and for justice. "I think that young people have a very unique ability [for] organizing, especially digitally, that a lot of older folks do not understand," she insists. "Understand that you are important and your voice is too, and don't let adults, especially, tell you otherwise."[29]

*Isra Hirsi (speaking), who is the daughter of Congresswoman Ilhan Omar of Minnesota, addresses a crowd in Washington, DC. Hirsi cofounded the US Youth Climate Strike, a group whose goals include bringing more kids of color into the climate movement.*

## Focusing on Youth

While Hirsi studied the effects of climate change on people in Los Angeles, Kevin Patel lived the reality. Patel is an Indian American, born in Los Angeles in 2001. During the years that Patel grew up, Los Angeles had the worst air pollution of any American city, according to the annual "State of the Air" report released by the American Lung Association. Patel blames the pollution for his ongoing medical problems; he sometimes requires hospitalization for heart palpitations that can cause shortness of breath, dizziness, chest pains, and feelings of panic. Patel said in 2019, "I recall spending most of my middle school years in and out of the hospital, with my parents always scared for my life. I now have to live with an irregular heartbeat for the rest of my life because of . . . the chemicals and smog that I was taking in."[30]

When Patel first looked into what triggered his heart palpitations, he was not thinking about air pollution. But after conducting research into the issue, he found that many in his community

## The Power of Youth

In 2020 Jamie Margolin published a how-to guide to student climate activism called *Youth to Power: Your Voice and How to Use It*. In the book Margolin interviews seventeen other young activists who provide a wealth of information about working for a healthier planet. As Margolin writes:

> The voices of young people are so powerful because we have the moral high ground. . . . Youth have nothing selfish to gain from our activism. There is hardly ever any monetary reward for being an activist, and any fame that comes along with it is rare and usually meager and fleeting.
>
> Youth don't speak out of a corrupt motive. We speak truth to power because we genuinely want change and to create a better world. And this is why the voices of youth are so pure and powerful, why they always have been, and why they always will be.
>
> Young people, despite our society conditioning us to follow the rules blindly, still have that knack for seeing right through the BS we are fed. The youth right now are the truth right now—and it's always been that way. Whether the adults who run our society admit to it or not.

Jamie Margolin, *Youth to Power: Your Voice and How to Use It*. New York: Hachette, 2020, pp. 19–20.

were harmed by the air-polluting gases released when fossil fuels are burned. "We were getting asthma, heart problems, and cancer because a lot of us live in communities where oil refineries are in our backyards,"[31] he said.

Patel discovered that air pollution was much worse in low-income communities of color, where factories, power plants, refineries, and major freeways were located. While air pollution and climate change might seem like two separate issues, Patel learned they are closely linked. When fossil fuels are burned, toxic pollutants such as nitrogen oxides, sulfur oxides, and carbon monoxide are released into the air, along with microscopic airborne particles called soot. During heat waves, when the air becomes stagnant, air pollution is trapped close to the ground. This pollution is responsible for several health problems, including respiratory infections, lung diseases, asthma, heart disease, liver cancer, and stroke. And according to an extensive 2019 study by the *New England Journal of Medicine*, air pollution kills more than 6 million people throughout the world every year.

Patel wanted to share this knowledge with other students. This led him to start the first high school environmental group in the South Los Angeles area in 2014. Patel soon expanded the club into a grassroots climate advocacy group called One Up Action. The mission of One Up Action is to increase the number of young leaders in the climate movement while providing marginalized youth with the educational and financial resources they need to tackle the climate crisis.

> "It's important that we have grassroots movements because [they] have to do with the communities that are being impacted by this crisis."[32]
>
> —Kevin Patel, founder of One Up Action

Patel says he views the future with a mixture of fear and hope that young people can help heal the planet. And he is a strong believer in community. "It's important that we have grassroots movements because grassroots movements have to do with the communities that are being impacted by this crisis," he states. "It's really important to mass mobilize the communities that are affected by these issues."[32]

# CHAPTER THREE

# The Teen Activist's Tool Kit

The sheer power of climate activism can be seen at street demonstrations when thousands march together and raise their voices as one. These large groups of people, focused on a strong, unified message, are difficult to ignore. But massive protests are the culmination of months of hard work by hundreds of individuals. And while the climate activists share common goals, individuals have their own reasons for joining the movement. Jamie Margolin calls this their "why," the main reason they became activists. Margolin believes that the first step for any prospective activists is to find their why:

> Before [you] get into any specifics of how one goes about community organizing and movement building, this is the foundational step of your changemaking journey: making it clear to yourself why you are an activist and what exactly it is you are fighting for. . . . Every single action you take going forward needs to serve your why. That's how you're going to be successful. That's how you're going to stick with it in the long run and make that change you have been striving for.[33]

Tokata Iron Eyes, a member of the Standing Rock Sioux tribe, knows exactly why she became an activist at age twelve. In 2016 a proposed oil pipeline project threatened the environment at the Dakota Standing Rock Indian Reservation where she lived. Iron Eyes described her journey into activism: "I was disrupted from being a kid to fight this pipeline. . . . It made me think about the other kids in my community who also cared but didn't have the resources to speak up. I decided to use mine to speak up and stand against the pipeline."[34]

## Where to Find Your Why

While activists like Iron Eyes find their voice when facing direct threats, you might my find yours in a less dramatic fashion. You might be inspired to become an activist after examining the effects of climate change on your family, friends, or wider community. Or you might need to take some time to think about why you want to commit to climate advocacy. The process does not need to be stressful, and it is not necessary to put your activism on pause until you come up with answers. As Margolin writes, "Have a good old honest chat with yourself. Be the annoying toddler who won't stop asking why. When you first try to give yourself shallow, superficial answers, keep probing 'why' until you can't go any deeper."[35]

> "I was disrupted from being a kid to fight [the Dakota] pipeline. . . . It made me think about the other kids in my community who also cared but didn't have the resources to speak up."[34]
>
> —Tokata Iron Eyes, climate activist

Transforming the world is not easy, and knowing why you are doing it will give you the inner strength you will need as an activist. Scientists say society needs to reduce greenhouse gas emissions by at least 80 percent by 2050. But this will require the passage of new laws that will change many entrenched aspects of society, including transportation, agriculture, home building, and power production. Climate activists are also asking millions of people to alter their habits and make personal and financial sacrifices for the common good. This includes buying electric cars, traveling less, and eating less meat (because meat production accounts for around 14 percent of greenhouse gas emissions). For activists the task will require commitment to the cause, hard work, patience, and passion.

## How to Make Your Case

While it might seem obvious that carbon emissions need to be quickly reduced to save the planet, climate activists are facing strong, well-organized opponents. And perhaps the biggest enemy of change is public apathy. According to a 2019 Quinnipiac University poll, around 42 percent of Americans are very concerned about climate change. This number has changed little since poll takers began asking the question in the 1990s. Activists who hope to counter apathy and opposition need to make strong, well-reasoned arguments founded in fact. This will require you to conduct intensive research to learn everything there is to

*Tokata Iron Eyes, a member of the Standing Rock Sioux tribe, became an activist at age twelve in 2016, when a proposed oil pipeline project threatened the environment at the Dakota Standing Rock Indian Reservation where she lived.*

know about climate change. While this might feel like an extra homework assignment, research is a vital tool that can be used to convince others that the issue merits action. Environmental activist Rob Greenfield explains the importance of being informed on your topic. He writes, "Really know your stuff so that you can inform the people you are looking to inform as well as go toe-to-toe with your opposition. . . . Use logic, rational, and concrete knowledge rather than emotion before going out and telling the world."[36]

> "Really know your stuff so that you can inform the people you are looking to inform as well as go toe-to-toe with your opposition."[36]
>
> —Rob Greenfield, environmental activist

An efficient way to conduct research is to find answers to questions reporters refer to as the five Ws: who, what, why, where, and when. Many activists begin their research on the internet— they just start googling. There are hundreds of factual newspaper articles, websites, environmental organizations, and government agency websites that explain the five Ws of climate change.

When conducting research, activists should avoid citing memes, conspiracy websites, biased blogs, and other questionable sources. In this era of intense political division, it is helpful to find neutral sources. For example, a study by a respected university about the effects of climate change on childhood asthma will have more credibility than a study funded by a large environmental organization with a political agenda.

Once you have a solid understanding of the issue, you can follow in the steps of Greta Thunberg. Before she tried to move the world, she practiced on her parents, who did not often think about climate change. Thunberg showed her parents documentaries about the changing climate and explained ways they could use less fossil fuel. According to Thunberg, "After a while, they started listening to what I actually said. That's when I realized I could make a difference."[37]

Thunberg said that practicing her arguments on her parents had several climate-friendly benefits. Her mother quit flying on airplanes, and her father became a vegetarian. Thunberg also learned how to make cohesive arguments to adults, which came in handy when she had to speak to the media, politicians, and business leaders.

## Volunteer for the Climate

While clearly explaining your cause is important, experience is the best teacher. Many activists begin their journey by joining an environmental club at school or an established organization. Margolin launched her activist career at age fifteen by working for mainstream politicians who supported a green agenda. She responded to an internet ad by the Washington State Democratic Party and was put to work as a volunteer. Her first job, making phone calls to get out the vote, initially made her very nervous. But things got easier. Margolin was soon made an intern and then advanced to the position of volunteer trainer. Soon Margolin was writing editorials, or op-eds, about the election for the local newspaper. As the only Latina in the office, she also worked as a Spanish-language translator. During this time she met senators, the governor of Washington, and presidential candidate Hillary Clinton. Margolin says her volunteering experience was amazing and catapulted her into a life-changing journey.

While volunteering might not transform your life, it is a great way to learn. Volunteers develop skills that are regularly used by climate activists, including patience, the ability to communicate clearly, teamwork, and leadership. Volunteers also learn how to raise money, manage an organization, promote events, analyze budgets, and raise awareness on social media.

Prospective activists need only to decide where they want to spend their time. Most environmental organizations allow volunteers to sign up on their websites, and helpers are not expected to have previous experience. All you have to do is show up, show that you are ready to learn, and perform the tasks that are given to you.

There are dozens of organizations that offer financial grants and monetary awards to small environmental groups. These can be used for education, community development, and environmental justice causes. For example, Patagonia, which makes environmentally sustainable clothing, provides grants from $5,000 to $20,000 to what it calls hyperlocal organizations. These grants are aimed at small grassroots groups that are working to address climate change. Another group, the Environmental Resources Management Foundation, provides grants of up to $150,000 to organizations that are committed to low-carbon development and to empowering women and girls.

A web search of environmental grants yields many foundations and groups that provide grants to small organizations seeking funding. Each organization has its own parameters for grant seekers, and some require detailed proposals. Those applying for grants should be able to clearly present their goals and explain how the funds will be used.

Most foundations only provide grants to nonprofit groups referred to as 501(c)(3) organizations. This status, which relates to a section of the tax code, means the group is legally recognized as a nonprofit organization. Most students recruit adults, including parents and lawyers, to help with the complex legalities required to set up a nonprofit 501(c)(3) organization.

## Starting Your Own Group

After you have honed your skills as a volunteer, you might want to set off on your own to address a local issue that is being ignored by other activists. Or perhaps you hope to become a national leader in the climate movement by building your own organization. This might prompt you to start something new—a movement or campaign that will respond to an unmet need.

While you might wish to tackle the issue alone, there is power in numbers. Surrounding yourself with like-minded people creates momentum, provides a forum for new ideas, and perhaps most important, adds an element of fun to an otherwise serious endeavor. If you are starting your own group, you will likely act as the organization's chair. In this role you will provide leadership for the group and work as a strategist to plan long-term project

*Social media is an ideal tool for finding like-minded individuals who support a cause. The Zero Hour group gained its momentum through a post on Instagram. Within a year Zero Hour had grown into a national organization. This July 2018 rally required careful planning by a large number of activists and a high level of coordination.*

goals. This job will require you to identify a problem you want to address and write out a clear plan for fixing the issue.

Once you have a solid plan, you need to spread the message and recruit a core group of people with similar beliefs. Social media is an ideal tool for finding like-minded individuals. Margolin started Zero Hour by posting a picture of a sign on Instagram for the Youth March on Washington. She explained her dream for Zero Hour in a post and was quickly contacted by a few people, who talked to more people. Margolin also sent out emails asking for advice from adult activists. Within a year Zero Hour had grown into a national organization.

You can follow the Zero Hour model by recruiting an inner circle that includes three to seven others who have a variety of talents. Activist groups need good speakers and people with writing skills and leadership abilities. The person with the best communication skills should work as a media coordinator. This person will shape the group's message into a set of consistent talking points. Media coordinators prepare statements for reporters and write op-eds for newspapers, magazines, and websites. The media coordinator is also in charge of managing social media, which can be central to the success of a campaign.

Most activist groups appoint a recruitment coordinator who works to attract supporters to the campaign. Recruitment coordinators put up flyers, speak to community groups, and search for like-minded activists on social media. If your group grows into a larger organization, the recruitment coordinator will manage inexperienced volunteers, who perform tasks such as making phone calls, creating signs, and maintaining websites.

The secretary of a group plays another important role. This job involves taking minutes of meetings, answering mail and emails, and managing whatever funds are raised by a group. Some groups have finance directors, who raise funds from the public and apply for grants from large organizations. In 2018 Madelaine Tew was the fifteen-year-old director of finance for Zero Hour. Tew secured a grant of $16,000 from the Common Sense Fund, which supports environmental change through education and promotion of carbon-free energy sources. She credits her success as a fund-raiser to classes she took in business and finance. These courses helped her navigate the complex granting process.

## Planning a Public Event

Most activists draw attention to their cause through public demonstrations like school walkouts and mass marches. As fourteen-year-old Alexandria Villaseñor said at the 2019 Global Climate

Strike in New York City, "When I go out and protest, it's one of the ways that I feel like I have a say in what's going to happen."[38] Villaseñor helped organize the strike, which was attended by around sixty thousand people. She understands that planning such events requires a large group of activists and a high level of coordination. Implementing an attention-grabbing event also takes time—up to a year for major actions. While it might seem like a daunting task, taking a methodical, step-by-step approach can make the planning process easier.

The first step in planning a street protest involves making a list. Identify the purpose of the event and the goals you hope to accomplish. Try to realistically estimate how many people you think will attend and pick a location where you want to hold the event. Activists hoping to hold a rally in a public park or march down a street will probably need to obtain a permit from the offices of the mayor, county clerk, or police department. City government websites usually provide information about the permitting process, but officials do not always cooperate. Authorities sometimes try to restrict routes or impose unrealistic noise levels on sound systems, claiming they are concerned about traffic control or public safety. But permits cannot be denied because an event is unpopular or controversial. Because it can take several months to book a venue and obtain a permit, this task needs to be addressed early in the planning process.

Long meetings are a major part of an activist's life, and as an event is coming together, groups need to brainstorm to complete a coherent plan of action. An event to-do list should include exactly what actions will be taken and who will oversee each task. The group needs to pick a venue for the protest and create an

> "When I go out and protest, it's one of the ways that I feel like I have a say in what's going to happen."[38]
>
> —Alexandria Villaseñor, Global Climate Strike organizer

## Organizing School Strikes

Greta Thunberg started an international climate movement in 2018 based on school strikes. The movement owes its success to thousands of individual student activists who organized climate strikes at their own schools as part of the Fridays for Future organization. On March 15, 2019, during the first global School Climate Strike, over 1 million people demonstrated in 125 countries. In Albany, New York, senior Audrea Din recalled learning about the upcoming strike on Twitter. "I only had a week to organize everything and I wasn't sure who was going to show up," she says. "I had to make a Go Fund Me and luckily I was able to raise $1,000 in just a couple days." Din used the money to set up a climate protest attended by two hundred students. She was later recruited as statewide New York coordinator for the September 2019 Global Climate Strike.

In Annapolis, Maryland, high school sophomore Kallan Benson helped bring together isolated supporters across the country for the first School Climate Strike. Benson used emails, phone calls, and Slack to organize kids in Florida, Washington, and elsewhere. As Benson explained, "We rely on social media to promote our strikes and get into the public psyche."

Quoted in Nadine Zylberberg, "Teens Around the World Are Skipping School to Strike for Climate Action," Medium, July 25, 2019. https://medium.com.

hour-by-hour schedule for the event. A list of prospective speakers and entertainers should be compiled, and a program coordinator should be appointed. The program coordinator will send out invitations early and work closely with speakers who agree to participate. The coordinator takes care of travel arrangements and acts as an emcee. Another person can work as a stage director and see that a stage, podium, and sound system are available for the rally. This job might involve hiring a local vendor that rents public address systems and other equipment. Most public events will require a security coordinator, who will work with volunteers to make sure everyone is safe. Large events might require hiring a professional security team. Students should consider recruiting other adults such as parents, relatives, and experienced organizers to help with security.

## Viral Activism

Attendance is the most important aspect of a rally, and the central task for an event planner is getting the crowds to show up. Fortunately, organizers have numerous social media tools at their disposal to spread the word. Before jumping on Twitter or Instagram, however, you need to create a brand for your campaign that will resonate with the public. Branding involves producing symbols, graphic emblems called logos, and catchphrases (which are short phrases or taglines that project your purpose). All are meant to work together to create a memorable public image for the group.

Once the brand has been established, it can be used to create a specific website for the event. This site will act as a base for your social media strategy, which will involve making posts about the event three or four times a week. Spread the word

*Social media has become a vital tool in event planning. One step in the organization of a mass demonstration is publicizing the event on a social media platform such as Facebook, Twitter, or Instagram.*

through tweets, emails, and posts on Facebook, Instagram, and elsewhere. Tie announcements of the event to relevant facts about climate change. For example, an email blast might link a recent extreme weather event to climate change and include photos and links to relevant scientific information.

> "Organizing a national strike is a lot like organizing a local rally/march, but . . . way more texts, email, decisions + zoom calls than I ever imagined humanly possible."[39]
>
> —Haven Coleman, climate activist

When inviting people to an event, provide a way for them to respond, or RSVP. Link the RSVP to your group's website or social media pages. You can send respondents texts to remind them of the protest and ask that they spread the word. You can also set up a number for SMS text messaging using a unique code that reflects your message. And postings also encourage supporters to use your group's hashtag as often as possible. Hashtags help promote a project to a national audience while effectively showing how much public support is behind the cause. And do not forget to ask for volunteers and donations.

## Finding Your Success

Climate activism is exhilarating, exhausting, and enlightening. Twelve-year-old climate activist Haven Coleman tweeted in 2019, "Organizing a national strike is a lot like organizing a local rally/ march, but op-eds, lots of op-eds and way more texts, email, decisions + zoom calls than I ever imagined humanly possible."[39] Creating a well-attended protest event can also bring feelings of joy and satisfaction. But implementing real change is an incremental process. That is why Margolin measures her success on a personal level. She asserts, "Being a successful youth activist does not look like only one thing. 'Successful' doesn't mean fame, it doesn't have to mean starting some massive organization, it doesn't mean writing a book. . . . It means whatever it is that you want it to be, whatever holds up your *why*."[40]

# CHAPTER FOUR

# Risks and Rights

In the late 1960s most youth-driven social movements were identified with charismatic leaders. Fred Hampton was widely known as a Black Panthers leader, while Abbie Hoffman was the face of the antiwar Yippie! movement. These figureheads were able to use their fame to raise money and publicize their movements. As leaders they also became the focus of public backlash and open harassment by the Federal Bureau of Investigation and other law enforcement authorities.

While the names of Hampton, Hoffman, and other leading activists are in history books, hundreds of others worked behind the scenes without recognition. The names and deeds of these activists have been forgotten by history. But social media has significantly changed this dynamic. Today anyone who promotes a cause on Twitter, Facebook, Instagram, or other platform can quickly gain thousands or even millions of followers. Some, like Greta Thunberg, become internationally famous for their climate advocacy after posting their personal protests on digital platforms.

Instant fame can be exhilarating, but there are risks associated with climate activism. After Thunberg gave a dramatic speech to the United Nations in 2019, she was targeted by trolls on social media. The haters did not usually address the issue of climate change. Instead, they attacked Thunberg and her family personally, posting violent threats, blatant lies, and fake photos. She was also criticized by TV commentators on Fox News, the governor of Kentucky, and even President Donald Trump. While critics accused Thunberg of being too young to understand the problem, her response made her appear more mature than those who harassed her:

> I honestly don't understand why adults would choose to spend their time mocking and threatening teenagers and children for promoting science, when they could do something good instead. I guess they must simply feel so threatened by us. But don't waste your time giving them any more attention. The world is waking up. Change is coming whether they like it or not.[41]

## Dealing with Harassment

Most teens have experience with online trolls. According to a 2018 Pew Research Center poll, 59 percent of American teenagers say they have been bullied or harassed online. And the number approaches 100 percent for climate activists and other social movement leaders. Anonymous harassers are most likely to target young women, especially those from diverse backgrounds. And according to a 2019 survey by the *Washington Post*, 68 percent of climate strike organizers and 58 percent of climate strike participants identified as female. But climate scientist Michael Mann says almost anyone who discusses the science of climate change in public can face coordinated attacks from opponents. Mann has learned to deal with the problem after twenty years, but he says those who go after young activists are shameful. "When they go after a sixteen-year-old girl like Greta—or other children—they expose themselves for the pond scum they truly are,"[42] Mann remarks.

"I honestly don't understand why adults would choose to spend their time mocking and threatening teenagers and children for promoting science."[41]

—Greta Thunberg, climate activist

While anonymous online threats can make someone feel powerless and afraid, there are ways to fight back. Cyberbullies only have the power you give them, and they try to make their targets feel isolated, alone, and helpless. If you have been attacked online, discuss your feelings with parents, trusted adults, and other activists. You might also want to report the harassment. If you are bullied on social media sites, submit a report to the platform. If you receive threats of bodily harm, contact the police; it is against the law to send threatening messages through electronic communication. Take a screenshot of the data and download your cell phone records to highlight the exact times the threats were made.

While it is important to take steps to reduce online threats, remember to keep things in perspective. As alarming as online harassment can be, do not give the trolls more power than they really have. Swedish sustainability professor Kimberly Nicholas maintains, "It's a tiny but vocal fraction of [an] already small minority that plays a disproportionate role in public discussions."[43]

## The Stress of Activism

Many young activists experience stress that goes beyond online harassment. Activism involves taking long conference calls, answering hundreds of emails, and constant texting. Activists working on big projects can experience burnout. The stress of activism can be particularly daunting for students balancing activism with the demands of homework, tests, keeping up grades, and filling out college applications. Student counselors warn that academic pressures, coupled with the stress of activism, can be unhealthy. Students need to learn to get off social media for a few days to regain a sense of balance in their lives.

Some activists set boundaries and strictly limit the time they spend working on their cause. You should never feel guilty about taking time off from your activism. Catch up with friends and family, take long nature hikes, spend time on hobbies, or go to a

movie or concert. Taking personal time to relax will improve your attitude and energy level when you return to your activism. Jamie Margolin writes:

> Let's change the world but THRIVE while doing it! The amount that you suffer for a cause does not equal how much you care about it or how successful it's going to be. Feeling like a martyr because you are working long hours with little to no pay, feeling like crap all the time, and barely scraping by for your cause is not wise. It doesn't make you superior, effective, or a hero; it just makes you sad. So don't be a martyr—be financially secure, be healthy, have fun, and prioritize your happiness.[44]

## Your Free Speech Rights

Activists who engage in civil disobedience that challenges school authorities, police, and powerful politicians face another element

## The Risk of Climate Anxiety

Climate activists spend their days focused on the ways humanity is harming the planet. They read reports that predict environmental disaster in 2030 or 2050, wondering how their lives will be disrupted if carbon emissions are not lowered. While climate activism can be empowering, there are mental health risks associated with focusing on negative environmental news. A 2017 report from the American Psychological Association (APA) calls this condition ecoanxiety. The toll that ecoanxiety can have on mental health includes feelings of depression, nervousness, helplessness, and panic. The APA report offers solutions for those suffering from ecoanxiety:

> People who bike and walk to work, school, appointments, and other activities not only reduce emissions and improve their physical health but also experience lower stress levels than car commuters. . . . Adolescents who actively commute to school show not only lower levels of perceived stress but also increased cardiovascular fitness, improved cognitive performance, and higher academic achievement. . . . Parks and green corridors have been connected to improved air quality and can increase mental well-being. More time spent interacting with nature has been shown to significantly lower stress levels and reduce stress related illness.

Susan Clayton Whitmore-Williams et al., "Mental Health and Our Changing Climate," American Psychological Association, 2017. www.apa.org.

of stress. Depending on their actions, they risk suspension from school and even arrest. One way to lower the anxiety associated with school strikes and climate marches is to know your legal rights.

The founders of the United States considered protest so important that they guaranteed the right in the First Amendment to the Constitution. This amendment says people have the right to speak freely, assemble peaceably, and petition the government to resolve their grievances. And in 1969 the Supreme Court ruled that there is no age limit on the First Amendment guarantee of free speech. The majority opinion, written by Justice Abe Fortas, has been quoted many times since it was written. Part of that opinion stated, "It can hardly be argued that . . . [students] shed

their constitutional rights to freedom of speech or expression at the schoolhouse gate."[45]

The court ruled that students can hand out flyers in common areas and speak out about their political beliefs. They can also wear clothing, such as T-shirts or hats, with slogans that express an opinion. There are limitations on these rights, however. If your school dress code bans all T-shirts or hats, you will be prevented from expressing your opinion by wearing these items. Student speech can also be restricted if it is disruptive to normal school activities. Student actions cannot hamper the work of teachers or classmates, nor can it prevent others from learning. These actions might include yelling out comments about a political issue during class or blocking hallways or doorways to protest.

While court cases have clarified student protest rights, teachers and school administrators might not always follow the law. School officials who disagree with a student's political message might view any protest as disruptive and try to ban it. And there are other limitations on free speech. The First Amendment was created by the founders to limit the power of the government. For this reason, court rulings on student free speech rights only apply to public schools. Private schools are not run by the government, and students who attend them are not necessarily offered the same protections.

## Know Your School Policies

The Fridays for Future school strike movement exposed another limitation on student rights. The law requires students to attend school. Those who stage a walkout or a strike are not legally protected from consequences. Unless there is an emergency, walking out of class in the middle of the school day or skipping school entirely is considered truancy. Courts have ruled that there is no constitutional right to violate truancy policies. Students can be disciplined for missing class—even if they are protesting issues that are important to them. The punishment for truancy differs by state and even between school districts. However, the penalty for truancy must adhere to a standard called content neutral. This means every student who skips school receives the same punishment; students cannot receive harsher punishment if they are expressing a political point of view. However, students who have committed previous offenses, whether or not they were related to political protest, can receive increased punishment.

Depending on where you live, you might be exempt from some truancy policies. In September 2019 over eight hundred climate strike events were held across the United States. In many small towns and rural areas, students were not allowed to skip school to join the strikes. But in Washington, DC, and Baltimore, students could strike if they first provided written permission slips from their parents. Those who joined a strike but did not provide permission slips were marked with an unexcused absence. In New York City; Los Angeles; Portland, Oregon; and elsewhere, millions of public school students were given unrestricted permission to skip school and join the strikes. As a result of the climate strikes, some teachers and politicians began referring to the protests as extracurricular activities. According to Vaughn Stewart, a member of the Maryland House of Delegates, "Civic engagement allows students to apply their classroom lessons to the real world. It is impossible for a classroom setting to perfectly simulate the experience of testifying in a hearing or organizing your classmates to march in the streets."[46]

> "It is impossible for a classroom setting to perfectly simulate the experience of . . . organizing your classmates to march in the streets."[46]
>
> —Vaughn Stewart, member of the Maryland House of Delegates

## Protect Your Rights on the Street

The right to assemble peacefully to express a political viewpoint is guaranteed to all Americans. And most street marches for the climate have been nonviolent. However, some actions by protesters cross the legal line, and police are quick to arrest those

It is important to know your legal rights if you are participating in protests. Bystanders are legally entitled to photograph demonstrations, and law enforcement authorities are not allowed to access photos or videos taken with a cell phone or camera without a warrant, even if a person is placed under arrest.

who violate the law. This was the situation in July 2019 when seventy protesters gathered in front of the *New York Times* headquarters in Manhattan. The activists demanded that all newspaper articles emphasize the dire climate situation by referring to it as a climate emergency. After the protesters blocked traffic with a die-in, lying in the street and pretending they were dead, sixty-six were handcuffed and arrested.

Legal experts say these activists crossed the line that separates legal from illegal behavior. Protesters do have the right to speak, chant, sing, and march in what are called public forums, including parks, city streets, and sidewalks. However, demonstrators like those in New York can be arrested if they block building entrances, detain pedestrians, or impede traffic.

There are other exceptions to the public forum rule. Protestors cannot demonstrate on private property unless the owner grants permission. Those who attempt to enter private property while expressing their political opinion can run into trouble with the law. For example, climate activists have been arrested for trespassing while protesting oil pipeline construction.

Courts have ruled that authorities can restrict the exercise of First Amendment rights according to strict time, place, and manner policies. This means officials can require protestors to obtain permits for large groups using a public park. They can also place limits on the hours when the protest can occur and on the volume of public address systems used for speeches. However, time, place, and manner restrictions must be content neutral, meaning the same rules apply to all public gatherings whether or not they are political.

While the rules and rights might seem contradictory or confusing, climate activists should use common sense. In August 2019 Zoe Schurman did not think about the consequences of her actions. She used spray paint to write "Strike For Our Futures!" on an outer wall of Seattle City Hall. Schurman, who was thirteen years old, was quickly arrested, handcuffed, and led away by police. As she later recalled, "I didn't really know what was going to happen, but they drove me to the precinct, and I was in a holding cell, actually, with my handcuffs still on. Then my parents picked me up."[47] A Seattle police spokesperson later said that anyone vandalizing public property would have received the same treatment as Schurman, who had to pay a fine.

A video of Schurman's arrest was posted on Twitter, and she was briefly internet famous. Reporters contacted her, and total strangers commented on her actions. Schurman regretted the graffiti arrest because it distracted from her legally protected actions. She organized the climate strike at her school and painted protest signs for others to carry. The day before the strike, Schurman gave a speech she wrote at a local press conference organized by a city council member.

## Dealing with Police

Those who shot video of Schurman's arrest were exercising a protected constitutional right. During public gatherings, activists are legally allowed to photograph or videotape the police, government authorities, and anyone else who is plainly visible.

This right was confirmed in 2017 by federal judge Thomas L. Ambro, who wrote, "Officers are public officials carrying out public functions, and the First Amendment requires them to [tolerate] bystanders recording their actions. This is vital to promote the access that fosters free discussion of governmental actions."[48]

The right to document the actions of authorities has exceptions. Police can legally order citizens to cease any activity that is interfering with law enforcement operations. Photographers may not place themselves between police and suspects or impede the movements of police officers.

> "Officers are public officials carrying out public functions, and the First Amendment requires them to [tolerate] bystanders recording their actions."[48]
>
> —Thomas L. Ambro, federal judge

If you are stopped or detained for taking photographs, never resist a police officer physically. Be polite and remain calm. The American Civil Liberties Union (ACLU) recommends that you ask only one question: "Am I free to go?" Until you ask to leave, your stop is considered voluntary and is legal. If the officer says no, then you are being detained. To detain you, the police must have a reasonable suspicion that you have committed a crime or are about to. Politely ask what crime you are suspected of committing. You can also remind the officer that taking photographs is a First Amendment right and does not constitute criminal behavior.

Courts have ruled that police may not scroll through the images on your cell phone or camera without a warrant, even if you are under arrest. Furthermore, officials cannot confiscate a smartphone or camera, demand that you delete photographs or videos, or delete your photos or videos on their own.

## Laws Target Climate Activists

After the terrorist attacks on the World Trade Center and the Pentagon on September 11, 2001, states quickly passed laws that designated facilities like dams and nuclear reactors as critical infrastructure. These laws included harsh new penalties aimed at terrorists who might, for example, attempt to sabotage them. Since 2017 politicians in at least seven states have targeted climate protesters by expanding the definition of critical infrastructure to include fossil fuel services. Protest techniques like trespassing at an oil fracking site or lying down in front of pipeline construction equipment were once viewed as misdemeanors. In Kentucky, North Dakota, Ohio, West Virginia, and elsewhere, these activities are now considered felonies that might include large fines and jail time.

Critical infrastructure laws vary from state to state, so it is important for climate activists to familiarize themselves with these rules. ACLU attorney Vera Eidelman explains, "This is a miscasting of protesters as economic terrorists and saboteurs. . . . Even if folks haven't been charged, the fact that these laws are on the books can seriously chill people and make them fearful of getting their voices out."

Quoted in Susie Cagle, "'Protesters as Terrorists': Growing Number of States Turn Anti-pipeline Activism into a Crime," *The Guardian* (Manchester, UK), July 8, 2019. www.theguardian.com.

While police are required to obey legal restrictions, they do not always follow the law, especially during protests when tensions escalate. Once again, common sense is called for. If you are threatened by aggressive authorities, it is in your best interest to leave the area immediately. Challenging baton-wielding police in riot gear is not going to advance your cause, but it could lead to personal injury or arrest. In less hostile situations it is best to follow the advice of the ACLU: "Know your rights and exercise them in a responsible way, and if you have to interact with law enforcement . . . or other authority figures, remember to be polite, calm, and clearly explain how your actions are within your constitutional rights."[49]

While the climate emergency might make you feel angry and upset, the First Amendment does not protect those who are intent on destroying property or causing physical harm to others. And acting out in rage tends to alienate those who might otherwise support your cause. In the past, environmental problems were solved when activists convinced a majority of Americans to support laws that protected the air, water, and endangered species.

Activism often requires protesters to defend their rights when confronted by school officials, the police, and other authorities. And those who wish to peacefully express their views are protected by more than half a century of court rulings. Those protections will be increasingly important as climate protests grow into an unstoppable movement in the years ahead. Thunberg said at a 2019 Extinction Rebellion protest in London, "We will never stop fighting, we will never stop fighting for this planet, and for ourselves, our futures, and for the futures of our children and our grandchildren."[50] Know your rights, fight the good fight, and do not give up hope for a better future.

# SOURCE NOTES

## Introduction: Climate Change Is Here and Now

1. Quoted in Marlene Cimons, "Meet Xiye Bastida, America's Greta Thunberg," PBS, September 19, 2019. www.pbs.org.
2. Quoted in Cimons, "Meet Xiye Bastida, America's Greta Thunberg."
3. Quoted in Sandra Laville and Jonathan Watts, "Across the Globe, Millions Join Biggest Climate Protest Ever," *The Guardian* (Manchester, UK), September 20, 2019. www.theguardian.com.
4. Quoted in Kristoffer Tigue, "Teen Activists Worldwide Prepare to Strike for Climate, Led by Greta Thunberg," Inside Climate News, September 19, 2019. https://insideclimatenews.org.
5. Quoted in Cimons, "Meet Xiye Bastida, America's Greta Thunberg."

## Chapter One: The Issue Is Climate Change

6. Quoted in Chris Mooney and Andrew Freedman, "Earth Is Now Losing 1.2 Trillion Tons of Ice Each Year. And It's Going to Get Worse," *Washington Post*, January 25, 2021. www.washingtonpost.com.
7. Quoted in Meteor Blades, "Blast from the Past—James Hansen, 1988," *Daily Kos* (blog), January 14, 2008. www.dailykos.com.
8. Quoted in Alan Buis, "Study Confirms Climate Models Are Getting Future Warming Projections Right," NASA, January 9, 2020. https://climate.nasa.gov.
9. Quoted in Philip Shabecoff, "E.P.A. Report Says Earth Will Heat Up Beginning in 1990's," *New York Times*, October 18, 1983, p. 1.
10. Intergovernmental Panel on Climate Change, "Executive Summary," Columbia University, 2008. www.ciesin.org.
11. Quoted in Anna M. Phillips and Evan Halper, "Biden Orders Sweeping Actions to Pause Energy Drilling and Fight Climate Change," *Los Angeles Times*, January 27, 2021. www.latimes.com.
12. Quoted in Cara Korte, "These Climate Activists Expect a Lot from President Biden and Aren't Afraid to Make That Clear," CBS News, January 20, 2020. www.cbsnews.com.

13. Quoted in Korte, "These Climate Activists Expect a Lot from President Biden and Aren't Afraid to Make That Clear."
14. Bill McKibben, "The Biden Administration's Landmark Day in the Fight for the Climate," *New Yorker*, January 28, 2021. www.new yorker.com.

## Chapter Two: The Activists
15. Kimberly Nicholas, "Climate Science Basics," 350.org, 2021. https://350.org.
16. Quoted in Carline Harrap, "Greta Thunberg's Dad: How Fighting for the Climate Changed My Daughter," The Local, December 19, 2019. www.thelocal.se.
17. Greta Thunberg, *No One Is Too Small to Make a Difference*. New York: Penguin, 2019, p. 3.
18. Thunberg, *No One Is Too Small to Make a Difference*, p. 32.
19. Quoted in Jamie Margolin, *Youth to Power: Your Voice and How to Use It.* New York: Hachette, 2020, p. 3.
20. Quoted in Sarah Sloat, "Jamie Margolin: Teen Activist Is a Leading Voice on the Climate Crisis," Inverse, April 12, 2019. www.inverse .com.
21. Quoted in Kristen Doerer, "Youth Climate Change Activists Marched on Washington, D.C.," *Teen Vogue*, July 22, 2018. www.teenvogue .com.
22. Quoted in Brooke Jarvis, "The Teenagers at the End of the World," *New York Times Magazine*, July 21, 2020. www.nytimes.com.
23. Quoted in Sloat, "Jamie Margolin."
24. Quoted in Jarvis, "The Teenagers at the End of the World."
25. Jamie Margolin, "Jamie Margolin's 2019 Congressional Testimony," House of Representatives, September 18, 2019. https://docs .house.gov.
26. Quoted in Rachel Hatzipanagos, "The Missing Message in Gen Z's Climate Activism," *Washington Post*, September 27, 2019. www .washingtonpost.com.
27. Quoted in *Teen Vogue*, "Isra Hirsi Talks to *Teen Vogue* About Organizing and Social Media," October 27, 2020. www.teenvogue.com.
28. Isra Hirsi, "A Learning and Growing Me," Medium, August 18, 2019. https://medium.com.
29. Quoted in *Teen Vogue*, "Isra Hirsi Talks to *Teen Vogue* About Organizing and Social Media."
30. Quoted in Nylah Burton, "Meet the Young Activists of Color Who Are Leading the Charge Against Climate Change," Vox, October 11, 2019. www.vox.com.

31. Quoted in Reem Sabha, "Kevin Patel: Environmental Justice Activist," *Medium*, November 12, 2019. https://medium.com.
32. Quoted in Sabha, "Kevin Patel."

## Chapter Three: The Teen Activist's Tool Kit
33. Margolin, *Youth to Power*, pp. 21–22.
34. Quoted in Margolin, *Youth to Power*, p. 230.
35. Margolin, *Youth to Power*, p. 24.
36. Rob Greenfield, "How to Create an Environmental Activism Campaign," Rob Greenfield (website), January 25, 2016. www.robgreenfield.org.
37. Quoted in Lydia Denworth, "Children Change Their Parents' Minds About Climate Change," *Scientific American*, May 6, 2019. www.scientificamerican.com.
38. Quoted in Carolyn Kormann, "New York's Original Teen-Age Climate Striker Welcomes a Global Movement," *New Yorker*, September 21, 2019. www.newyorker.com.
39. Quoted in Alejandra Borunda, "These Young Activists Are Striking to Save Their Planet from Climate Change," *National Geographic*, March 13, 2019. www.nationalgeographic.com.
40. Margolin, *Youth to Power*, p. 179.

## Chapter Four: Risks and Rights
41. Quoted in Rebecca Falconer, "Greta Thunberg Addresses Climate Deniers' Attacks and Tweets," *Axios*, September 26, 2019. www.axios.com.
42. Quoted in Elizabeth Weise, "Online Haters Are Targeting Greta Thunberg with Conspiracy Theories and Fake Photos," *USA Today*, October 2, 2019. www.usatoday.com.
43. Quoted in Weise, "Online Haters Are Targeting Greta Thunberg with Conspiracy Theories and Fake Photos."
44. Margolin, *Youth to Power*, p. 203.
45. Abe Fortas, "Tinker v. Des Moines Independent Community School District," Cornell Law School, 2017. www.law.cornell.edu.
46. Quoted in whmacken2013, "Skipping School to Protest. No Excuses," ThinkingOregon, January 4, 2020. https://thinkingoregon.org.
47. Quoted in Matt M. McKnight, "'We Should Be Able to Have a Normal Childhood': Seattle's Young Climate Activists on Why They Strike," Crosscut, September 20, 2019. https://crosscut.com.
48. Quoted in Matt Ford, "A Major Victory for the Right to Record Police," *The Atlantic*, July 7, 2017. www.theatlantic.com.
49. American Civil Liberties Union of Tennessee, "Stand Up/Speak Up: A Guide for Youth Activists," 2015. www.aclu-tn.org.
50. Thunberg, *No One Is Too Small to Make a Difference*, p. 68.

# WHERE TO GO FOR IDEAS AND INSPIRATION

## Books

Jamie Bastedo, *Protectors of the Planet: Environmental Trailblazers from 7 to 97*. Markham, ON: Red Deer, 2020.

Stuart Kallen, *Teen Guide to Student Activism*. San Diego: ReferencePoint, 2019.

Jamie Margolin, *Youth to Power: Your Voice and How to Use It*. New York: Hachette, 2020.

KaeLyn Rich, *Girls Resist! A Guide to Activism, Leadership, and Starting a Revolution*. Philadelphia: Quirk, 2018.

Greta Thunberg, *No One Is Too Small to Make a Difference*. New York: Penguin, 2019.

## Organizations and Other Websites

### American Civil Liberties Union (ACLU)
www.aclu.org

The ACLU works to defend individual rights guaranteed in the Constitution. The organization's website features numerous articles with legal information about student rights, free speech, racial justice, privacy rights, and LGBTQ rights.

### Climate Reality Project
www.climaterealityproject.org

The Climate Reality Project was founded by former vice president Al Gore to mobilize more than nineteen thousand Climate Reality Leaders, who push for practical clean energy policies across the United States and elsewhere.

### Earth Guardians
www.earthguardians.org

Earth Guardians is a student organization made up of activists, artists, and musicians dedicated to empowering young people to take over as leaders of the environmental movement. The group's website features comprehensive information about environmental issues and ongoing campaigns.

## 17 Environmental Grants to Fund Your Project
https://blog.temboo.com/environmental-grants

This article provides information to those seeking monetary grants from foundations and organizations committed to climate advocacy. Each listing provides details about who is eligible and how much money is available, along with links to grant applications.

## 350.org
https://350.org

This organization, founded by climate activist Bill McKibben, is focused on banning oil, coal, and gas production. The website offers resources for organizing and training activists, including art, graphics, and resource materials.

## US Youth Climate Strike
https://strikewithus.org

US Youth Climate Strike focuses on racial, environmental, indigenous, and immigrant justice and how these issues relate to fighting climate change. The group's youth committee provides monetary grants to small student environmental groups.

## News Articles

Brooke Jarvis, "The Teenagers at the End of the World," *New York Times Magazine*, July 21, 2020. www.nytimes.com.

Carolyn Kormann, "New York's Original Teen-Age Climate Striker Welcomes a Global Movement," *New Yorker*, September 21, 2019. www.newyorker.com.

Sarah Sloat, "Jamie Margolin: Teen Activist Is a Leading Voice on the Climate Crisis," Inverse, April 12, 2019. www.inverse.com.

Michael Patrick F. Smith, "The First Step Is Admitting You Have a Problem," *New York Times*, February 5, 2021. www.nytimes.com.

Susan Clayton Whitmore-Williams et al., "Mental Health and Our Changing Climate," American Psychological Association, 2017. www.apa.org.

## Documentaries

Bonni Cohen and Jon Shenk, dirs., *An Inconvenient Sequel: Truth to Power*. Los Angeles: Paramount Pictures, 2017.

Leila Conners, dir., *Ice on Fire*. Schenectady: Appian Way, 2019.

Nathan Grossman, dir., *I Am Greta*. Los Angeles: B-Reel Films, 2020.

## Apps

### Earth-Now

https://climate.nasa.gov/earth-apps

NASA's Earth Now app displays global data in real time including $CO_2$ levels, daily temperatures, dust storms, hurricanes, and sea level variations. Color-coded maps indicate the strength and level of various environmental conditions.

### Ecosia

www.ecosia.org

When users search the web using this free app they see advertisements that generate income for Ecosia. Eighty percent of the ad profits are donated to groups that focus on planting trees in South America, Africa, and elsewhere. Ecosia protects user privacy; it does not track users or sell data to advertisers.

# INDEX

# PICTURE CREDITS

# ABOUT THE AUTHOR

Stuart A. Kallen is the author of more than 350 nonfiction books for children and young adults. He has written on topics ranging from the theory of relativity to the art of electronic dance music. In addition, Kallen has written award-winning children's videos and television scripts. In his spare time he is a singer, songwriter, and guitarist in San Diego.